Introduction to Parkinson's Disease

Dr. Robert Fekete

Copyright

Preface

The purpose of this book is to provide an introduction to the world of Parkinson's disease to patients and caregivers. I endeavored to provide a historical background to currently utilized treatments and also to discuss aspects of the disease that may not be readily apparent from the initial visit with a physician.

While the information in the book has been sourced from current best practices, please note that medical inflation changes rapidly. Before engaging in a treatment, please discuss the latest known side effects with your physician and read the package insert of the medication you are about to take. This book does not provide specific treatment recommendations and reading this book or making suggestions to the text does not start a physician patient relationship.

Suggested additions and corrections can be sent to my office at 19 Bradhurst Ave, Hawthorne, NY 10532 and will become a part of the next edition. I would like to see this book grow as a comprehensive resource.

I hope that the information here will allow you to enter into more productive discussions along the entire care team consisting of patient, caregiver, other family members, and allied health professionals such as physical therapists.

Sincerely,
Dr. Robert Fekete

Table of Contents

History

James Parkinson provided the comprehensive description of the disease that bears his name in 1817 via his Essay on Shaking Palsy. The disease is also called paralysis agitans, which encompasses two important clinical findings - difficulty initiating movement combined with periods of being unable to control movement such as festination or hurried gait.

Hurried or festinating gait was identified prior to James Parkinson's essay but not in conjunction to all of the other symptoms he described. This gait was termed sclerotyrbe festinans by the physician Sauvages. James Parkinson tied all of the currently known symptoms together, except for rigidity (muscle stiffness), which was emphasized by Charcot. Jules Froment further investigated the clinical finding of rigidity. The Froment maneuver by which rigidity increased with movement of the opposite limb remains important in modern clinical examination. Please see the next chapter for a discussion of these classic symptoms.

The disorder involves the loss of dopamine producing pigmented nerve cells, which are located in the substantia nigra. These cells have projections to part of the brain called basal ganglia which can be compared to a transmission in a car as the function often basal ganglia is to modulate movements. The analogy can be drawn further if you have seen an inexperienced driver improperly use a manual transmission. The car will have difficulty starting and lurch forward in a jerky manner instead of a smooth start and acceleration. A similar effect occurs with gait in more advanced Parkinson's disease.

Initially, researchers discounted dopamine as just being a precursor to norepinephrine. Arvid Carlsson worked with dopamine depleting agents such as reserpine and postulated that dopamine is important in motor function based on animal experiments. Oleg Hornykiewicz's experiments with post-mortem brain tissue slices of Parkinson's patients showed less dopamine in the striatum (which receives connections from the substantia nigra). Subsequent research by Arvid Carlsson, George Cotzias, and Melvin Yahr demonstrated improvement of motor symptoms in patients treated with levodopa, a precursor of dopamine. Levodopa that converts to dopamine outside of the brain contributes to nausea, which was profound in early formulations. Carbidopa blocks the conversion of levodopa into dopamine outside the brain and significantly reduces nausea. The trademark for carbidopa levodopa combination is Sinemet,[1] derived from sin emesis or without nausea/vomiting.

Prior to the use of levodopa, anticholinergic medication was the primary method of treatment, with Charcot using scopolamine in 1867. This class of medications reduces the amount of acetylcholine. It is believed that this allows remaining dopamine to be more effective. Unfortunately, acetylcholine is used for memory storage and other thinking processes. Blocking it can lead to confusion and even hallucinations, which are reversible once the medications have stopped. Anticholinergics are used more in young onset PD and in patients without thinking and memory problems due to these issues.

In the 1970s, Dr. Paul Marsden's group in London investigated the use of dopamine agonists to treat PD. The first agent identified was bromocriptine, which had significant side effects and is currently used for endocrinological purposes and not PD treatment. Pergolide was taken off the market for cardiological issues. Currently available dopamine agonists are pramipexole, ropinirole, rotigotine patch, and apomorphine injection.

---

[1] Sinemet is a registered trademark of Merck & Co., Inc.

Current research is focusing on why neurodegeneration occurs in the first place. There are multiple theories which include toxic (such as MPTP causing direct toxicity to substantia nigra in a few cases of accidental ingestion), lysosomal ("garbage disposal" structures in the cell which may fail), genetic predisposition (for example one allele of the gene for Gaucher disease versus the better known LRRK2 gene), and protein misfolding whereby abnormal proteins cause normal proteins to change into abnormal conformations in a domino-like reaction. Much more research is needed before any of these theories or a combination become more definitive.

Classic features of the disease

The classic motor features of the disease can be summarized by the acronym TRAP, which means tremor, rigidity, akinesia, and postural instability. The most classic tremor is rest tremor and is typically described as a pill rolling tremor. The physician Gowers described it as fingers beating on a small drum. In addition, there may also be a tremor with posture and reaching movements (called action tremor). The presence of action tremor sometimes complicates the diagnosis of mild early Parkinson's disease versus essential tremor, the prototypical action tremor disorder. Numerous peer reviewed publications have been written on this issue, with the general consensus being that if action tremor developed 5 years before the onset of Parkinson's disease, it is most likely from co-existing essential tremor. Some patients with essential tremor may have rest tremor, which further complicates the issue.

Rigidity is examined by the physician and is classified according to degrees. Some patients with mild rigidity may have it appear only with the Froment maneuver, which is why your physician may ask to move the contralateral arm during examination.

Akinesia, or bradykinesia, means slowness of movement. This is often the symptom that family members and patients notice the most. They will describe handwriting becoming slower and also smaller over time. Movements of the entire body become slower and less fluid. Speech is of lower volume. Classically, the physician will ask you to perform repetitive movements to judge the speed and amplitude. Bradykinesia translates to gait as well, with the classic PD patient having a stooped posture with decreased arm swing usually on one side.

As the disease progresses over time, gait and balance disturbance becomes more prominent. Especially in milder disease, your physician may perform the pull test, where you are observed to see how many half-steps you need to regain your balance. The utility of this maneuver decreases when the patient is displaying obvious balance disturbance including wide based gait and requires a cane or other assistive device to ambulate.

Freezing of gait occurs later in the disease and may be triggered by tight spaces such as elevator doors, small bathrooms, and attempts to get onto an escalator. Distraction can also exacerbate freezing of gait. Various non-pharmacological remedies that focus on visual and auditory cues are used to counteract this phenomenon. These include marching or performing exaggerated steps, possibly while listening to marching music, placing visual cues on the floor in the bathroom, or using a mobile visual cue such as a cane or walker that projects a laser light onto the floor.

Pharmacologic strategies in young onset versus late onset Parkinson's disease

Young onset disease is defined as having motor symptoms prior to the age of 50. The disease is more slowly progressive than classic PD. There is an increased risk of dyskinesia from the disease itself and from the body's adaptation to the use of dopaminergic medication for decades.

Monoamine oxidase inhibitors of which the best known is rasagiline are often used in treatment of mild early PD. They allow existing dopamine to work better because they block the enzyme which breaks it down. Typically, there is mild improvement of symptoms usually felt the most in the morning.

Dr. Olanow and colleagues performed a reanalysis of the STRIDE-PD study, where they noted risk factors for the development of motor fluctuations. These factors include young age at onset and higher levodopa dose. Levodopa sparing strategies are used as much as possible (especially in young onset PD) in order to delay the onset of dyskinesia, but these strategies have their own drawbacks. Anti-cholinergic medication is very useful, but is limited by reversible cognitive dysfunction leading to confusion, rarely even hallucination, and also rare possibility of urinary retention. Dopamine agonists are very useful medications, but the patient and caregivers have to watch out for sleep attacks as well as compulsive behaviors. Amantadine, originally a medication used for the flu, can have a beneficial effect on PD symptoms, especially tremor, but has its own side effects including anxiety and livedo reticularis, which is a reversible lacy rash usually on the lower extremities.

After side effects occur and the levodopa sparing strategy is no longer useful, levodopa needs to be introduced. There are currently more choices, with both immediate and longer lasting formulation of levodopa available. You will discuss which of these options is right for you with your physician.

In late onset patients, usage of anti-cholinergic medication and dopamine agonists may be more likely to include cognitive and other side effects. Hence, these medications are used more carefully after discussing possible risk factors for side effects - such as pre-existing sleep apnea in a patient who is considering dopamine agonist. In both young onset and late onset PD, Deep Brain Stimulation (DBS) surgery is carefully considered after thoughtful discussion of the goals of the patient and possible risks and benefits of the surgery.

Motor fluctuations

Over time, the "honeymoon period" of Parkinson's disease medications lasting well until the next dose during the daytime makes way for motor fluctuations. These include periods of "ON" time mixed with "OFF" time during which the medication stops working before the next dose becomes effective. Patients may have a variety of "OFF" symptoms which could include freezing of gait, worsening of bradykinesia, muscle spasms (dystonia) which especially seem to occur in the toes, and return of rest tremor. In many patients, at the time they develop OFF symptoms, they also develop dyskinesia, which are involuntary "dancing" type movements. The period of time during which these occur is called "ON with dyskinesia." Movement disorders specialists have a variety of ways to reduce motor fluctuations, which include, changing levodopa from immediate to longer lasting formulation, changing the frequency of levodopa administration, adding in amantadine to reduce dyskinesia, or using a rescue medication such as apomorphine during "OFF" periods. It is important for both the patient and their caregiver to carefully document what they typical day looks like, with medication dosage times and times when they enter "OFF" as well as "ON with dyskinesia" periods. Your physician needs to know with which dose the problems occur in order to be able to provide a cogent solution to modifying the pharmacotherapy. Even in clinical trials where patients are specifically coached in using a diary with appropriate nomenclature, it may be difficult to understand the terms precisely. If you have difficulty with classifying a particular movement, please ask your caregiver to describe it and perhaps take a cell phone video of it to show to your physician. That way, your physician will be able to accurately classify whether you are complaining of an "OFF" symptom such as tremor which typically needs more medication versus "ON with dyskinesia." It will be much easier to communicate your symptoms to the physician during the visit once these concepts are clearly understood and described. If the "OFF" symptoms typically occur during a particular time of day, your physician will be able to move relevant medication administration times appropriately.

Constipation

Research is showing that the nervous system of the gastrointestinal tract is affected early in the disease course. Lewy bodies, the microscopic features associated with neurodegeneration in PD, have been found in colonoscopy biopsy specimens of patients taken prior to the PD clinical diagnosis.

Natural remedies for constipation include prune juice. Patients usually escalate to over the counter remedies such as Colace and sennosides. If these are ineffective, prescription strength laxatives are prescribed. In some cases, the damage to the gastrointestinal system includes the stomach's pacemaker as well. In patients with advanced PD, it means that the stomach may not always process medications when they are taken and patients may have a 'dose failure' of an oral medication. I have had patients in this situation tell me that they have to eat a small meal to stimulate or jump start the stomach in order to process a medication they have taken. Alternatives to oral medications can be considered in these settings. These include patch form of medication such as rotigotine, injectable medication such as apomorphine, and deep brain stimulation surgery (DBS).

Sleep disorders

REM behavior disorder (RBD) can commonly precede the development of Parkinson's disease. It involves acting out dreams. Dreams occur during REM (rapid eye movement) sleep. Normally, the body is paralyzed during REM sleep. In RBD, the body is not paralyzed while dreaming. As a result, patients can talk and move during sleep. In patients who have violent dreams, the movements can include punching. In rare cases, patients can fall out of bed. Hence, it would be recommended not to keep nightstands with sharp edges next to the bed if someone is suspected of having RBD. An over the counter remedy for RBD is melatonin, up to 9 mg prior to going to sleep.

Restless legs syndrome (RLS) is also associated with Parkinson's disease. Of note, medications such as dopamine agonists are FDA approved for the treatment of both RLS and Parkinson's disease. RLS is defined as an uncomfortable sensation (usually inside the legs) associated with urge to move which is relieved by movement. Hence, a patient with RLS may come up with remedies such as pacing around the bedroom or massaging the legs prior to going to sleep. Some medications and over the counter remedies can actually bring out the symptoms of RLS. These include anti-histaminergic medications that are marketed for insomnia. Anemia can also exacerbate the symptoms of RLS. Spouses of patients with RLS report seeing limb movements during sleep. These are reported as muscle jerks and are termed periodic limb movements of sleep. Unlike movements of RBD, these do not imitate voluntary limb movements.

Compulsive behaviors

Compulsions can be insidious. For example, after stopping a dopamine agonist and starting levodopa, a patient may realize that she was buying unnecessary clothing. This behavior may seem normal to a patient while on the dopamine agonist. It is thought that the D3 dopamine receptor is most involved in compulsive behaviors. Apomorphine, which is a D1 and D2 agonist, should theoretically be less likely to induce compulsive behavior than the classic oral D2 and D3 agonists ropinirole and pramipexole.

Exercise

Exercise is an important part of PD treatment. A review in the Movement Disorders Journal found improvement in gait, balance as well as in general well being. My patients pursue walking, tai-chi, and LSVT BIG.

LSVT BIG is a set of exercises specifically designed to improve movements in PD.

LSVT LOUD is a speech therapy designed to teach a patient to speak with higher voice volume.

Other types of physical, speech, and occupational therapy can be selected depending on a patient's individual needs.

Hallucinations

Hallucinations usually involve small animals and people. They may start as benign, for example seeing children playing in the yard. Over time they start to involve more distressing and nefarious visions. For example, the children playing in the head can become adults who are damaging the patient's favorite trees and flowers. Some describe people whizzing by in the hallway and quickly disappearing. In most cases, the apparitions vanish when the person tries to touch them. The most persistent hallucination that was reported to me was a case of a face in a bedsheet crease. After pulling on the bed sheet, it vanished but quickly reformed on another portion of the bed sheet.

The hallucinations can be distressing to not just the patient but also the caregiver. It may be very distressing for the caregiver to have to explain every day that there isn't anyone in the yard.

Delusions are even more destructive to patient interactions with caregivers and friends. Patients may have a delusion that a caregiver is stealing from them and wake up every night to look their spouse's belongings to look for this stolen item. Delusions of infidelity from a spouse who is the primary caregiver may lead to nursing home placement as the relationship is eroded by false accusations. Delusions of someone breaking into the home can lead to checking behaviors, such as having extensive routines to ensure that all doors and windows in the house are securely locked. In some cases, patients can interpret flashing light hallucinations as a laser light being pointed at them and barricade themselves behind their bed, which also leads to nursing home placement.

Elevated levels of dopamine lead to more hallucinations and abnormal thoughts. Neuroleptic medications which are dopamine receptor blockers are used in both the neurological and psychiatric worlds to control hallucinations. Quetiapine and clozapine are used but effectiveness is limited. I had difficulty completely eliminating hallucinations and delusions with these medications. It is generally thought that dopamine agonists are more likely to trigger hallucinations than levodopa itself. In my practice, I had more success with reducing dopamine agonist and levodopa intake than with using neuroleptics. Of course, limiting pharmacological treatment will increase motor symptoms of Parkinson's disease. In one patient, I found a compromise with rotigotine patch 1 mg per day, which is a very low dose of dopamine agonist, but the only medication that could be tolerated without disabling hallucinations.

Pimavanserin is a novel serotonin inverse agonist medication which promises to significantly advance the treatment of psychosis in Parkinson's disease. It is hoped that the serotonin based mechanism will not cause exacerbation of motor symptoms which is the main concern with dopamine receptor blocking agents such as quetiapine.

Parkinson's disease dementia and diffuse Lewy body disease

The classic definition is that if someone has symptoms of dementia first or within one year of developing motor symptoms of Parkinson's disease, they have diffuse Lewy body disease (DLB). If dementia symptoms occur after one year, the disorder is called Parkinson's disease dementia (PDD). On brain specimens, pathologists can not distinguish between PDD and DLB. Given this and other similarities between the two disorders, there is a movement underway to reclassify these disorders.

Frontotemporal dementia, multiple system atrophy, progressive supranuclear palsy, and corticobasal syndrome

In some cases, patients that see me for Parkinson's disease may actually have atypical parkinsonism. Some of the clues for atypical parkinsonism may be difficulty with downward gaze, which is a clue for progressive supranuclear palsy, limb apraxia - unable to move or even recognize one's fingers or even the entire hand in corticobasal syndrome, and profound behavioral disturbance as well as speech disturbance in frontotemporal dementia.

Multiple system atrophy (MSA) has a well known cerebellar (with incoordination as the primary symptoms) as well as parkinsonian variant, with both variants requiring profound disturbance of the autonomic nervous system (which regulates blood pressure and other "automatic" body functions). It may be very difficult to separate idiopathic PD with autonomic dysfunction from the parkinsonian variant of MSA.

Consultation with a movement disorders specialist is encouraged especially if atypical parkinsonism is considered in the differential diagnosis.

Notes

Index

Author Bio

Dr. Robert Fekete is a Phi Beta Kappa graduate of Johns Hopkins University. He received his medical degree from NYU School of Medicine in 2005. After residency training at North Shore - LIJ Health System, he completed a two year Movement Disorders fellowship at Baylor College of Medicine.

In addition to his peer-reviewed publications in neurology, Dr. Fekete is a contributor to Medlink.